# better together*

*This book is best read together, grownup and kid.

 akidsco.com

a kids
book
about

# a kids book about

# BEING A SCIENTIST

by Christopher Reddy

**A Kids Co.**
**Editor** Emma Wolf
**Designer** Rick DeLucco
**Creative Director** Rick DeLucco
**Studio Manager** Kenya Feldes
**Sales Director** Melanie Wilkins
**Head of Books** Jennifer Goldstein
**CEO and Founder** Jelani Memory

**DK**
**Senior Production Editor** Jennifer Murray
**Senior Production Controller** Louise Minihane
**Senior Acquisitions Editor** Katy Flint
**Acquisitions Project Editor** Sara Forster
**Managing Art Editor** Vicky Short
**Managing Director, Licensing** Mark Searle

First American edition, 2025
Published in the United States by DK Publishing, 1745 Broadway, 20th Floor,
New York, NY 10019

First published in Great Britain in 2025 by
Dorling Kindersley Limited, 20 Vauxhall Bridge Road, London SW1V 2SA
A Penguin Random House Company

The authorised representative in the EEA is
Dorling Kindersley Verlag GmbH. Arnulfstr. 124, 80636 Munich, Germany

A catalog record for this book is available from the Library of Congress.
A CIP catalogue record for this book is available from the British Library.
ISBN: 978-0-2417-4313-3

DK books are available at special discounts when purchased in bulk for sales
promotions, premiums, fund-raising, or education use. For details, contact:
DK Publishing Special Markets, 1745 Broadway, 20th Floor, New York, NY 10019
SpecialSales@dk.com

Printed and bound in China
**www.dk.com**
**akidsco.com**

**MIX**
Paper | Supporting
responsible forestry
FSC™ C018179

This book was made with Forest
Stewardship Council™ certified
paper – one small step in DK's
commitment to a sustainable future.
Learn more at **www.dk.com/uk/**
**information/sustainability**

To my children: William, Elliot, and Flynn,
and to aspiring scientists around the world.

# Intro
## for grownups

**K**ids are natural learning machines and endlessly curious: How do planes fly? Why is the ocean salty? Where does the wind come from? They long to understand the world and their place in it. And too often, as they grow up, their curiosity can dwindle, and they stop wondering.

There's a name for kids who grow up and never stop wondering—they're called scientists.

We can encourage budding scientists as they explore how things work, imagine what will happen next, and contribute research which will make life better for everyone. As a scientist, I love that I never run out of perplexing problems to solve and paths to explore—some deadends, others full of promise—which explain how the world works. I wrote *A Kids Book About Being a Scientist* to share the thrill and reward of asking questions and trying to figure out the answers.

Hi there! My name is Chris.

# AND I'M A SCIENTIST.

When I say that...what do you imagine my job is like? What comes to mind?

Whatever that is...
let's put it aside for now.

As a scientist, I get to...

EXPLORE,

ASK QUESTIONS,

GO ON WALKS AND COLLECT THINGS TO STUDY,

LEARN FROM OTHERS,

SHARE AND TEST MY IDEAS AND EXPERIENCES,

AND TELL STORIES.

Did you expect me to say any of that?

# SOUNDS PRETTY COOL, RIGHT?

If you look up a definition of a scientist, you'll find something like:

# A PERSON WHO DOES EXPERIMENTS TO ADD OR REFINE EXISTING KNOWLEDGE.

But there's more to it than that.

Being a scientist gives
you the opportunity to be
# CURIOUS

# AND LEARN

about the world around us.

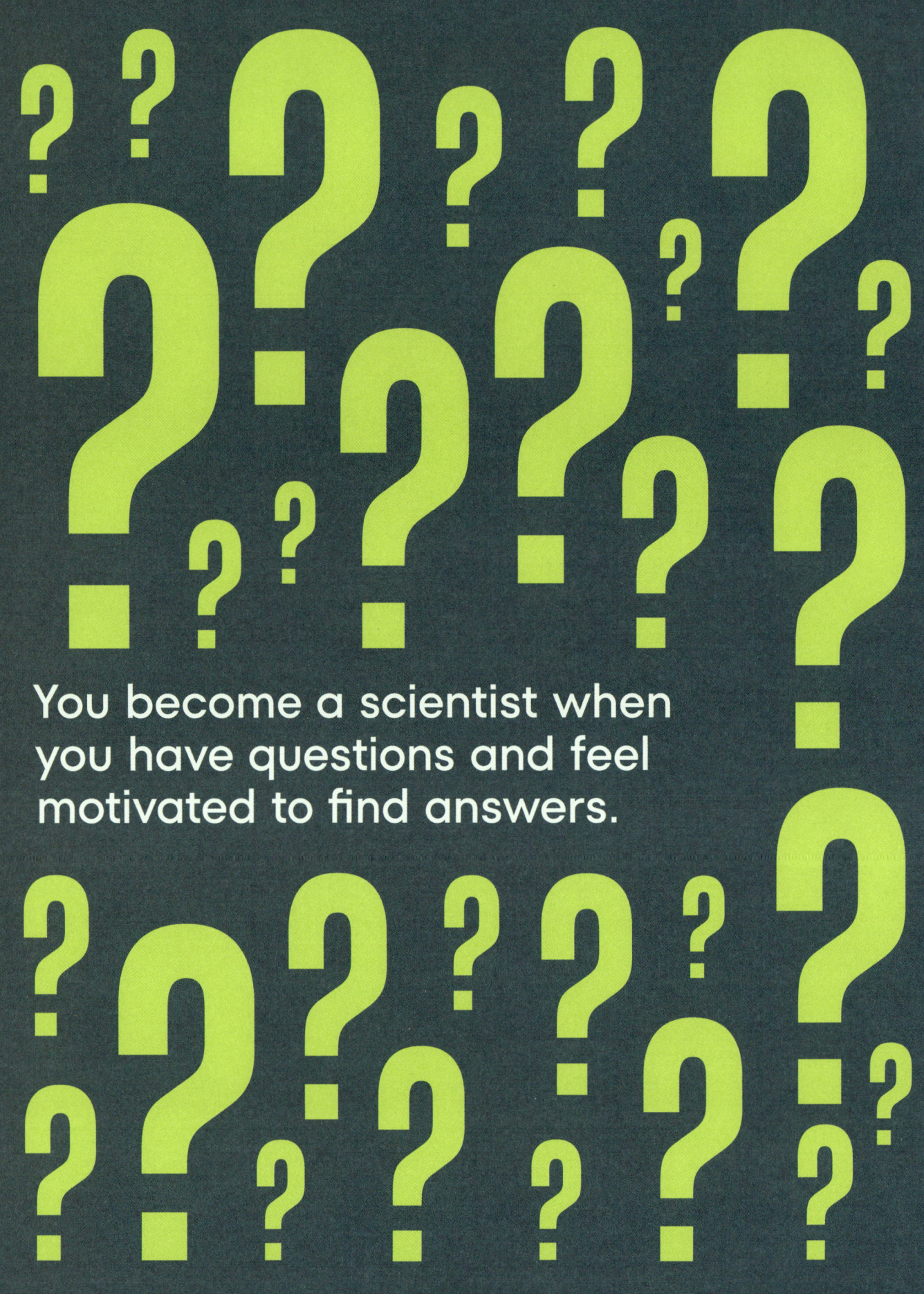

You become a scientist when you have questions and feel motivated to find answers.

And you can do this anywhere, any time, with whatever you have around you!

It's about what matters to you, what you discover, and what you want to share.

# EVERYONE CAN BE A SCIENTIST, EVERY DAY!

I think of science as a giant jigsaw puzzle.

There are so many pieces that require patience and curiosity to find where they fit in.

You can work on the puzzle alone, or you can bring together a team of people to solve it.

Sometimes, you don't find the "right" piece. Science isn't perfect (and neither are scientists!).

Science is about trying things, seeing how they work, and making changes to see what happens.

Being a scientist is about the journey—you're building knowledge along the way.

# SO, WHAT MAKES A GOOD SCIENTIST?

# CURIOSITY!

It is important to be interested
in solving problems that
really matter to you.

Scientists ask "Why?" and "How?" a lot, and they're willing to keep asking while they search for answers.

# PATIENCE

is a big part of it.

A lot of the questions scientists want
to answer can't be solved quickly.

A dedicated scientist is willing to stick with their work, even when days feel long or boring—and even when they begin to doubt themself.

# PERSEVERANCE.

You may not succeed,
but you did not fail.

Scientists don't always get things right

the first time around...

or the second...

or even the third.

It's expected and normal to make mistakes. A scientist learns from those mistakes instead of giving up.

And when you do find what you've been looking for...

# THAT IS THE BEST FEELING!

Like scoring a goal in a soccer game, or giving the perfect speech in front of your class.

It is so rewarding to see the results of your hard work, and it makes all the time invested completely worth it.

It is also fun to tell your friends, family, and teachers what you've learned!

## AND HERE'S A BIG ONE:

# GOOD SCIENTISTS

**AND** **WILLING TO CH**

# ARE OPEN-MINDED ANGE THEIR MINDS.

Sometimes, that's not
an easy thing to do!

But when we dive into big questions, all kinds of things come to the surface.

Scientists are excited to learn from those around them and will change direction when the facts tell them to.

# BEING A SCIENTIST IS SO AWESOME.

So, how can you be a scientist in your everyday life?

**I'M SO GLAD YOU ASKED!**

When I visit schools through my work, the students and I investigate and explore the playground.

**Then, we all tell stories about what we see:**

*Why do puddles form here and there (but not in other places) after a rain storm?*

*Why is the grass greener on one side of the playground?*

*Why did that tree fall during the last storm?*

*Why is only 1 garbage can full?*

*What makes it sunny, rather than cloudy?*

# WE STUDY WHAT WE FIND AND WORK TO UNDERSTAND WHAT WE CAN LEARN FROM EACH OTHER.

And you can do the same thing at home or at school!

You can go out with your grownups and friends and see what you find.*

WHAT'S ON THE GROUND OUTSIDE?

HOW ABOUT IN THE PARK?

OR AT YOUR SCHOOL?

WHAT DO YOU SEE?

WHAT DO YOU HEAR?

WHAT DO YOU FEEL?

WHAT QUESTIONS DO YOU HAVE
ABOUT WHAT YOU'RE EXPERIENCING?

Scientists stay curious, ask questions, and are excited to search for answers!

AND THERE'S SOMETHING **IMPORTANT** I WANT YOU TO KNOW.

A lot of people don't believe they have what it takes to be a scientist.

I hope by now you
know that isn't true.

But, in case it hasn't
totally sunk in yet...

# THAT IS ➡

# NOTTRUE!

I really mean it when I say anyone can be a scientist.

Your life is your laboratory!

# WHAT EXCITES YOU?

## WHAT ARE YOU CURIOUS ABOUT?

### WHAT PROBLEMS DO YOU SEE THAT YOU WANT TO FIND SOLUTIONS FOR?

I hope you keep working to find the answers to these questions.

From one scientist to another...

# ENJOY
# THE RIDE!

# Outro
## for grownups

Being a scientist is fun, challenging, and exciting. Kids don't need a laboratory and fancy equipment—they only need to think, ask questions, and try to solve them. Working alone or with a team, they can pick the project and stop whenever they want. They can pick the time and the place. Celebrate with them when they figure something out or share what they have learned.

And here's something sneaky: studying the world around me helps me better understand others and myself. When I am having a bad day or feel frustrated, I use my scientist brain, try to learn why I am upset, and then look for solutions.

Countless studies point to the benefits of pursuing open and engaging conversations between kids and grownups. Talking science is a perfect vehicle to build meaningful relationships with your kids. And watch out, grownups—you might catch the "science bug" too!

# About The Author

Christopher Reddy (he/him) is a globally recognized oceanographer at the Woods Hole Oceanographic Institution on Cape Cod. He is curious about how the ocean responds to uninvited guests like plastic and oil spills. Christopher uses his findings to help design safe and sustainable drinking straws and replace petroleum products with natural waxes in sunblock.

Christopher credits his grandfather for kickstarting his lifelong interest in understanding how things work, and his mom for impromptu chemistry demonstrations in the kitchen. Christopher is a proud graduate of Rhode Island's public schools, from kindergarten through PhD, and he can tell you the name of every science teacher he has ever had.

He constantly asks, "I wonder why...?" and lives in Woods Hole, Massachusetts, with his wife, Bryce, and their children, William, Elliot, and Flynn.

 @chris-reddy-science           christopherreddy.com

# Made to empower.

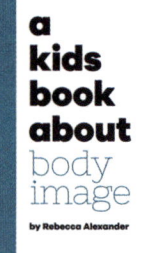

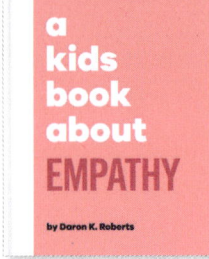
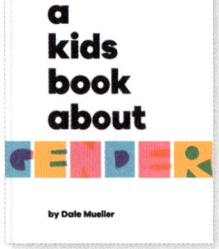

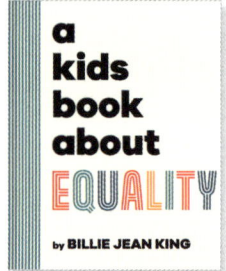

## Discover more at akidsco.com